Pierce the Skies

Review
by Vincent Dublado, *Reader's Favorite*
Review Rating: ★★★★★

Dar Bagby combines her passion for birds and poetry in *Pierce the Skies: Poems about Birds in Michigan's Upper Peninsula*. From the American bittern to the yellow-bellied sapsucker, she pays tribute to her bird-watching experience within Michigan's Upper Peninsula where she resides. She did her best to include most of the bird species that inhabit her hometown, and most of the birds in this collection are the ones that particularly inspired her. While she humbly admits that she is not a math wizard, it would be hard for any lover of poetry to overlook her careful attention to the exact numbering of her syllables and the beat and pace of her lines that strengthen her words. Aside from her iambic pentameters, her attempts at different meters create rhythms that create different sound and visual effects, generating a mood that puts you into a natural background with birds flying, tweeting, or nesting. In one of her poems, Pileated Woodpecker, the bird illustrates its natural behavior captured in a combination of short lines that would make you stop to watch the bird drill a hole in the trunk of a robust tree.

Pierce the Skies may well have been written to show Ms. Bagby's vision of nature and how it represents the formation of her ornithological poetry. From age four where she heard her father's bird poetry reading for the first time, this anthology is an inspiring revisiting of that moment in her life. As for us readers, behind the imagery of every bird represented herein echoes a suggestion of romantic and euphoric interaction with the feathered creatures that inhabit our skies. This work is reminiscent of the tradition for which Romantic literature has been widely known, as it abounds in ornithological themes. For anyone who appreciates the harmony between man and nature, *Pierce the Skies* should be on your reading list.

Pierce the Skies

Poems about Birds in Michigan's Upper Peninsula

Dar Bagby

Copyright 2020 by Dar Bagby

ALL RIGHTS RESERVED. No part of this book may be reproduced or transmitted in any form by any means, electronic or mechanical, including photocopying, scanning and recording, or by any information storage and retrieval system, except as may be expressly permitted in writing by the publisher. (darbagby@gmail.com)

Front cover photo by Kunal Baroth
Interior artwork by Connie M. Thompson
Manufactured and written in the United States of America

Published by Dar Bagby

First Edition
Printed in the United States of America

ISBN 978-1-0879-0173-2

Library of Congress Control Number: 2020914067

The opinions in this book are the author's opinions only and are freely offered. They are not meant to be offensively portrayed.

DEDICATION

…to all the birders with whom I love sharing birding experiences.

ACKNOWLEDGEMENTS

Thanks to my friend and fellow birder, Connie Thompson, who is also my graphic designer. *Your expertise in paginating this book and preparing it for publication is unsurpassed. I give you all the credit for making it so attractive.*

Thank you to my husband, Ken, for his patience while I was totally absorbed in my writing efforts. *You can't imagine how much easier that made my job.*

TABLE OF CONTENTS

Introduction 1
Accipiters, Buteos, Northern Harrier 50
American Bittern 74
American Goldfinch 65
American Kestrel, Merlin, Peregrine Falcon 95
American Robin 20
American White Pelican 47
Bald Eagle 23
Baltimore (Northern) Oriole 81
Barn Swallow 38
Belted Kingfisher 78
Black-capped Chickadee 103
Blue Jay 88
Brown Creeper 45
Bufflehead 36
Canada Goose 26
Cedar Waxwing 70
Common Loon 77
Common Merganser, Hooded Merganser 52
Common Raven 91
Common Yellowthroat 97
Dark-eyed Junco 24
Downy Woodpecker, Hairy Woodpecker 41
Evening Grosbeak, Pine Grosbeak,
   Rose-Breasted Grosbeak 84

Flycatchers (Tyrant, Empidonax) 68
Golden-crowned Kinglet, Ruby-crowned Kinglet 58
Gray Jay 63
Great Blue Heron 72
Great Gray Owl 54
Gull and Tern 19
Horned Lark 110
Indigo Bunting 33
Killdeer 113
Lapland Longspur 5
Mallard 94
Mourning Dove 99
Northern Flicker 49
Northern Pintail 8
Osprey 43
Pied-billed Grebe 105
Pileated Woodpecker 61
Piping Plover 28
Red-breasted Nuthatch, White-breasted Nuthatch 76
Ring-necked Pheasant 35
Ruby-throated Hummingbird 15
Ruffed Grouse 101
Sandhill Crane 29
Snow Bunting 9
Snowy Owl 32
Song Sparrow 67
Trumpeter Swan 10
Turkey Vulture 90
Veery 30
Warblers (in spring) 86
White-throated Sparrow 107
Wild Turkey 111
Wood Duck 13
Yellow-bellied Sapsucker 56

INTRODUCTION

On a warm Saturday afternoon when I was four years old, my mother delegated to my father the responsibility of making certain I took a nap. He spread a blanket on the ground in our back yard and proceeded to tell me that we were going to look for birds, a sport in which, to that point, I had never participated. So we both lay down, eyes to the sky, and looked for feathered figures passing overhead, my father probably figuring I'd quickly be in Slumberland. It was on that very day that my father granted me the pleasure of hearing my first poem about birds:
>*Little birdy in the sky,*
>*Dropping whitewash in my eye,*
>*Gee, I'm glad that cows can't fly.*

Of course, to a four-year-old, that was pure delight, and I rewarded my father with giggles and the request to "say it again." Which he did. Several times. And I became an avid birdwatcher, though I cannot, in all honesty, say it was due, even in part, to that episode.

Now I have taken it upon myself to redeem the

dignity lost on my first exposure to bird poetry and so have written my own poems about many of the birds of Michigan's Upper Peninsula, where I reside. The majority of the poems in this collection are based on my actual experiences. There are, of course, countless other birds of the U.P. which I did not include; the ones I chose to write about are the ones that most inspired me to do so.

Though I have never claimed to be a math whiz, numbers do fascinate me. You'll find this evident in my poetry, especially in the precise number of syllables per line in my "rhythmic consistency" offerings. In addition to the familiar iambic pentameter, I have included multiple meters in multiple types of poems.

Far and away, my favorite bird poem of all time is this one by Robert Frost:

I have wished a bird would fly away
And not sing by my house all day.

Have clapped my hands at him from the door
When it seemed as if I could bear no more.

The fault must partly have been in me;
The bird was not to blame for his key.

And, of course, there must be something wrong
In wanting to silence any song.

I can only hope I have done myself justice with my participation in Mr. Frost's art form.

LAPLAND LONGSPUR
(Calcarius lapponicus)

A spring walk on a dreary day
Along the shores of Whitefish Bay,

I had binoculars in hand
To see what might appear, unplanned.

I lay on sand, head on a log,
And looked up at the lifting fog.

I know I won't see much today.
I wondered just how long I'd stay.

I raised my binocs to my eyes
And peered into the milky skies,

Then heard a noise just to my right
And slowly turned to see a sight

That shocked me. So I stayed dead still,
And what I saw was such a thrill:

A Lapland longspur, new for me
(A bird for my life-list, you see).

It stood not six feet from me there.
And motionless, it met my stare.

Its eyes looked tired; it had just flown
For miles and miles from parts unknown

On its way north, without a rest,
To find a mate and build a nest.

But it had stopped along our shore.
Had this bird flown this path before?

It didn't move; it only stared.
I'm guessing it was really scared.

Or maybe not; instead, it might
Have been exhausted from its flight.

I quickly tried to memorize
Its silhouette, its shape and size,

I only got a tiny glance
Before it bravely took the chance

To tilt its head and notice me
As if it suddenly could see

That I was foreign to its world.
And so—its wings abruptly curled—

It sprang into the cloudy sky
And left me wondering just why

I'd chosen to be in that spot
At that choice moment. Had I not

Been there that day to see a sight
So wonderful, I'd have no right

To think back on that bird and smile
And be glad that I stayed a while.

NORTHERN PINTAIL
(Anas acuta)

What a handsome male!
    Slender neck of white,
Head of choc'late brown,
    Eyes of darkest night,
Long, extended tail,
    Pointed wings in flight.

Female's smaller though,
    Body mottled brown.
Neck is pale and tan,
    Darker on the crown.
Belly's white, bill's dark,
    Chest conceals soft down.

Abundant and widespread,
    Found on ponds and lakes,
Open areas
    Attract both hens and drakes.
But when winter comes,
    They avoid snowflakes.

SNOW BUNTING
(Plectrophenax nivalis)

A flash of white against dark pines,
The whirlwind hurtles up from the cold ground
And becomes transparent in the winter sky.

TRUMPETER SWAN
(Cygnus buccinator)

"We've lost our last cygnet," she said to the cob.
"Then we will start over. The bear will not rob
Us again. I'll make certain. I'll be more attentive."
"You are not to blame. We must be more inventive.
We'll follow the shoreline instead of the pathways
And keep to ourselves during bright sunny days."
She stretched out her long neck and sang to the sky.
And he followed suit; in duet they did cry.

They went about tidying up the nest site.
They raised a new brood, which they cared for just right.
All of the cygnets grew large and were healthy.
That is, all save one; that fox was so stealthy.
"We couldn't have done any better," he told her.
"Our little one lagged, and the fox simply stole her."
She sighed and said, "Yes, dear, I know that's the story.
I just can't stop seeing it, bloody and gory,
Our little one screaming and begging for help,

And us flapping 'round, making yelp after yelp."

She lowered her head underwater and dabbled.
The cob, all the while muttered short squawks
 and babbled.
When Mama Swan's head came back up, the
 cob said,
"We still have five youngsters. We can't lose
 our heads.
They'll make us proud parents. Now stay on
 your guard."
"I promise I will," Mama said. "But it's hard."

At that very moment the pair spied a bobcat.
It hid in the rushes; so quietly it sat.
It stared at the cygnets who played near
 the shoreline.
Its whiskers were twitching; the muscles of its spine
Contracted; its eyes grew enormous; it hugged
 the ground.
They knew there'd be no chance to stop it
 mid-bound.

The swans opened up with loud hissing,
 heads bobbing.
Cat never moved; its heart had to be throbbing.
And then they exploded. With powerful wings beats
They battered the rushes and voiced squeals and
 loud bleats.

The bobcat jerked, turned, and flipped into the air;
It landed on all fours then ran from the pair.

"We did it, my darling! We scared her away.
Our cygnets will live one more glorious day!"

WOOD DUCK
(Aix sponsa)

I had a laugh at its expense.
It happened years ago
When I was at an aviary
Taking in the show
Of waterfowl that floated by;
One, a male wood duck.
I guess the woman next to me
Assumed I was a schmuck.

(Now, this is hard for me to tell
Without a major snort:
I'll do my best to self-contain,
But I won't leave you short
Of entertainment at this point.
I know that you'll laugh, too.)

I said, "Oh look. A wood duck,"
And the woman came unglued.
She whirled around and leered at me.
I leaned back on my heel.

She said to me with such disdain,
"Oh no, my dear. It's real!"

RUBY-THROATED HUMMINGBIRD
(Archilochus colubris)

Cute, he is,
    tiny feet,
    whirring wings.
His throat patch
    ruby red
    reflecting
    in sunlight.

Figure eights,
    backing up,
    hovering.
Dive displays
    when he courts.
Squeaky chirps.
Rapid *tiks*
    made with wings.
Faint whining
    tail feathers
    when he dives.

His mate builds
> thimble-sized
> lichen nest;
> soft lining—
> spider's silk.

Two wee eggs.
Eighteen to
> twenty-two
> days to fledge.

One brood? Two?

Nighttime sleep,
> heartbeat slows;
> in torpor.

Michigan
> enemies:
> sharp-shinned hawks,
> kestrels, too,
> blue jays, cats,
> bats, chipmunks,
> squirrels, and
> big praying
> mantises.

Nectar from
> tubular
> flowering
> plants and trees.

Preference:
> pink and red
> and orange.

Tongues are long,
> extend far.

Minerals,
> vitamins,
> and protein—
> arthropods.

Some tree sap,
> sugar-rich,
> taken from
> wells of the
> sapsuckers.

Migrates south:
> Panama,
> West Indies,
> Mexico,
> Florida.

Doubles his
> fat reserves
> by one gram
> in prep for
> Gulf crossing.

Fattens up
> once again

    in the spring
    so he can
    return to
    the U.P.,
    early May.

Certainly
    welcome here!

GULL AND TERN

Flap and float, gull and tern.
    Rise and fall.
    Live and learn.

AMERICAN ROBIN
Michigan's state bird
(Turdis migratorius)

Though you represent my state,
You're not a bird I cherish.
I've banned you from my "favorites" list.
You strike me as quite garish,

Not because your chest is red
Nor 'cause you cheerily sing;
Mostly it's the way you act;
You're such a ding-a-ling.

Run and hop. Stop. Cock your head
And listen for a meal:
Earthworms, soft invertebrates.
They offer great appeal.

But as for me, who works so hard
To keep my flower gardens
Well stocked with worms to aerate soil,
I offer you no pardons.

First you pounce and then you pull;
You do your very best
To expose worm—life on the line—
To feed your brood in nest.

Worm clings and screams (though silently)
And stretches to full length.
But usually you win. Alas,
Worm cannot match your strength.

As if that isn't bad enough,
You hunt with more insistence
By scratching up my garden mulch;
It offers no resistance.

You seem to have no conscience when
You're tearing up my labors.
In fact, you dare me to approach you,
Rattling my sabers.

Off you fly with gullet full.
You'll transport what you've taken
And offer up the spoils to all
Those tiny mouths a-gapin'

And begging you to feed the one
Whose red maw's most attentive.
Instinctual skills are right on track;
You need not be inventive.

But when my brain convinces me
To angrily condemn you,
I must consider what I do
To fill my own child's menu.

Without a thought I offer up
Some cow or fish or chick part.
I can't condemn your parenting;
We share a mother's heart.

BALD EAGLE
(Haliaeetus leucocephalus)

Regal bird
Symbol of a nation
Icon of liberty

Soaring without wing beats
Eyes looking down on me
Without realizing my tribulations

Sculpture of you
In gold, bronze, or silver
Atop a flagpole

You alone
Are permitted to stand almost imposingly
Above the stars and stripes

DARK-EYED JUNCO
(Junco hyemalis)

I see you, little junco, hopping on the snow
Underneath my feeders. Will you stay or go?
Minding your own business, looking for a seed
Either missed or dropped; exactly what you need.

A blue jay lands above you and emits a squall.
You flit into a conifer. With "chips" you call
Your aggravation to the one who ends your search.
But till he eats his fill, you won't leave your perch.

I know there are more than one subspecies of you,
Slate-colored and *Oregon*, to name only two.
Up here in the U.P., slate-colored's what we see:
Gray back and white underparts. You show
  up commonly.

All of you share outer tail feathers of bright white
Which flash open now and then, mostly when
  in flight.
Your bill is thick and strong, its color is light pink.

Your head is rounded, tail is long—handsome,
  I think.

You're one of the sparrows which also grace
  our lawn
In spring, summer, and fall. I surely do count on
Seeing you in winter when the others depart.
You show up when it's cold and always warm
  my hcart.

CANADA GOOSE
(Branta canadensis)

The golf courses hate you. (I think you know why.)
And beach-goers secretly wish you would die.
You're noisy and dirty. You chase people, too.
You tolerate humans because they feed you.
And predators? They have been mostly removed
Through natural processes. And you've approved.
The urban environment now is your home;
And mostly, you're no longer eager to roam.

But here, every spring you return once again;
I welcome you back like a long-lost best friend.
Your honking and cackling are joy to my ears.
I love how you've lifted me up through the years.
When I see you winging and hear your loud voice,
I know that I've certainly made the right choice
About living here in the Eastern U.P.,
As you provide great entertainment for me.

I have quite the honor of watching your style
The way it was meant to be; that makes me smile.

You two are partners; you're mated forever.
You work hand-in-hand (wing-in-wing
  or whatever.)
Your goslings are precious when swimming in line
Between you and your mate. It's such a good sign;
I love watching all of the things that you do,
And I can appreciate you being you.

PIPING PLOVER
(Charadrius melodus)

Few are left—
Making comeback,
Numbers increasing.

Whitefish Point
Closes beaches
When the plovers nest.

Protection
Has been helping.
Birds from other spots

Sighted there
With new banding
Methods in process.

Hate to think
The world might lose
Something so special.

SANDHILL CRANE
(Grus canadensis)

I hear the cranes. Do you?
My head turns t'ward the sky.
I squint into the sun
To see them passing by.
Their heads are stretched straight out,
Their feet lag far behind,
Their wings push through the air.
They seem to pay no mind
To those of us below.
They don't acknowledge me
When they are overhead,
But fly expectantly
To light in yonder field
With hopes of finding corn.
They'll mate and eat their fill,
And when their colts are born,
I'll chuckle to myself
When they come marching by
With fuzzy orbs on stilts
Till those young learn to fly.

VEERY
(Catharus fuscescens)

The veery is a kind of thrush;
It isn't much to look at.
The thing that sets this bird apart
Is its bright voice, and just that.

Birds don't use their larynx
To vocalize their sounds.
Instead, they have a syrinx
That creates ups and downs

For their vocalizations.
And some can master more
Than only one at any time.
The veerys meet that score.

They sing a rolling series of
Graceful descending notes
That roll and spill and tumble
From their impressive throats.

It's difficult to miss them
When they begin to trill.
Their voices echo through the woods.
I never get my fill

Of hearing them. I must admit,
It's something far beyond
The normal sound of any bird.
And I am truly fond

Of walking through the forest
And list'ning for their song.
I'm certain, when I hear it,
That I'm where I belong.

SNOWY OWL
(Nyctea scandiaca)

Atop a utility pole you perch, looking down on the
   world…yours.
You are aged; only a speckle of dark feathers
   soils your pristine whiteness.
I stare in awe at your prowess.

Your head swivels, and your piercing eyes
   meet mine
As if to ask, "How dare you invade my very soul?"
I avert my gaze,
Confused because that reaction seems not of my
   own accord.

INDIGO BUNTING
(Passerina cyanea)

One day in late spring
I came to a stop sign at the end of a dirt road.
Ahead of me was a field that had been allowed to
 go fallow.
Perched on the stems of overgrowth were black,
 sparrow-sized birds
I could not identify.

I watched them sway in the breeze at the bidding of
 the wind,
Which was coming from behind me,
So I could hear no identifying call or song.
I decided to move on.

I began to turn the corner, and as a result,
The direction of the sun lit up that field
In a shade of bright blue,
The likes of which I'd never before seen.
I blinked a few times before realizing

It was not the field that was blue, but instead,
The birds.

I stopped dead still, right in the middle of the turn,
And gazed upon nearly a hundred indigo buntings,
All vibrantly blue,
Refracting the sunlight within the structure of their
  non-blue feathers.
My mouth hung open.

RING-NECKED PHEASANT
(Phasianus colchicus)

Pheasant's a game bird like turkey and duck,
Grouse, dove, and goose; they're all out of luck
When people are hungry for meat on their plates.
These birds must accept that they may meet
  their fates
By thrilling a hunter who's eager to shoot
A meal that is feathered and tastes good, to boot.
I'm one of those people who likes to eat meat,
I'm not above shooting it. I think it's sweet
To provide my own sustenance; it makes me proud
To know I can feed myself—and I'm allowed.
It isn't so much that I enjoy the kill.
But slow-roasted pheasant's a connoisseur's thrill.
Whether I roast it or grill it or fry,
It makes my mouth water…I simply can't lie.
And so, Mr. Pheasant (or Mrs,) beware:
You're apt to discover I'm your worst nightmare.

BUFFLEHEAD spirit duck
(*Bucephala albeola*)

Handsome
>Color: striking black and white male,
>Shiny green and purple on head
>With large white patch behind the eye

Small
>Rival green-winged teal as smallest.
>Fit nest holes of Northern flicker.

Flock
>Small groups; one remains a sentry;
>Sky-bound predators are big threats.

Dive
>They have high metabolism,
>Dive almost continuously
>Searching for insects, plants, fish eggs.

Monogamous
>Females seek the same breeding site.

Nests: small cavities in aspens;
No competition with large birds.

Prolific
Clutch sizes: six to eleven.
Nest success rate is very high.
One day after last egg hatches
Brood leaps from the nest cavity.
Fledge: fifty to fifty-five days.

Threats
Humans, weasels, minks, other birds,
Yet numbers tend to stay constant.

Migrate
One of last to leave breeding grounds;
They are quite punctual migrants.

BARN SWALLOW
(Hirundo rustica)

Just to the south of Whitefish Point
A short way 'round the bend,
There's a harbor where I go,
A place I'd recommend
To any birder who has time
To sit and rest awhile
And watch for hawks and terns and gulls.
They're sure to make you smile.

The harbor walls are old but strong;
They're made from coffer dam.
I stopped there only yesterday
To eat the cheese and ham
That I had purchased from a store
Not very far away.
I thought I'd eat and maybe nap,
But that plan went astray.

I opened up the window
And took a deep breath in.

The fresh lake air filled up my lungs;
My lips curled in a grin.
The breeze was just this side of cold.
I welcomed it and thought,
It's been so hot for so long now,
I'm glad I'm in this spot.

But…

There'd been a hatch of stable flies
Just recently, I s'pose.
In seconds they were in the truck
And bit my shins and toes.
Reluctantly I scrambled,
Quickly shutting myself in.
I'd have to forfeit the fresh air.
Imagine my chagrin.

I settled in to eat my lunch
Without that cool fresh air.
The thrill was gone; my spirits dashed.
It just seemed so unfair,
Almost like being punished
For something I'd not done.
I tried to make the best of it;
I tried to make it fun.

Then all at once I found myself
Surrounded by a flock

Of small dark birds with rufous chests.
I stopped, and I took stock…
Barn swallows! I so love to watch
Them play and chase and dart
Around the puddles in the drive
As if they're mobile art.

DOWNY WOODPECKER
(Picoides pubescens)
HAIRY WOODPECKER
(P. villosus)

(Dis poem is dedicated ta all da Yoopers who've asked me 'bote dese two birds.)

'Bote dese birds, I'm often asked, "Da small one's a
   baby, right?"
Da answer's, "No, dat's not da case." (Den I git ta
   sound bright!)
"Der are some major diff'rences, most obvious—
   da size.
Da downy's just six inches tall; da hairy's nine,
   youse guys.
Da side uh da downy's tail has dots; da hairy, he
   ain't got 'em.
Da hairy's bill is longer; check it ote next time ya
   spot 'im.
Da males have red spots on der heads. An'
   zygodactyl feet;

Dat means two toes dat go each way: two front,
 two back. It's neat!
Der calls are diff'rent, too," says I. "Da downy's
 call goes down,
Da hairy's stays on da same pitch." (An' dis is
 where dey frown.)
"I didn't know ya knew so much 'bote dis bird
 watchin' stuff."
I look 'em in da eye an' smile; I tink dat says
 enough.

OSPREY fish hawk, fish eagle, sea hawk
(Pandion haliaetus)

You do not hide behind your dark mask.
Fly proudly over water. Spot prey.
Enter the water feet-first,
Talons closing, piercing scales and skin
Of oblivious fish that do not make a habit of
  looking up,
But get *carried* up out of their familiar world
To sate another, though not by choice.

You do not hide behind your dark mask.
Call, neither insipid nor raucous, announces...
No...shouts your claim of mate and territory.
"She is mine for life!" and "I have chosen this place
To teach my fledglings to find sustenance."
Those that succeed will teach their own one day,
They who must rely on cooperation of ecosystems.

You do not hide behind your dark mask,
You, who takes advantage of habitat world-wide,
Not just my little corner of it.

You, who builds your home in full view of
   your enemies,
Your bravado, that of a mythical Grecian king—
   your namesake.
You, whose fossils appear alongside the bones
Of earliest *Homo sapiens*, Pleistocene to Holocene.

You do not hide behind your dark mask.

BROWN CREEPER
(Certhia americana)

Driving U.P. backroads on a windless sunny day,
A movement caught my eye, and so I slowed.
I knew no leaves were moving; it must have
 been alive.
I stared at where I'd seen it by the road.
I sat for what seemed ages; I couldn't let it go.
(It was determination overload.)

I was, at last, rewarded when I spotted a small bird
That shifted upward on a maple tree.
It blended with the tree's bark; I barely made it out.
But it, no doubt, could easily see me.
It had a bright white eyebrow above a dark
 black eye.
Its tail was long and fanned out feathery.

I slowly raised my binocs and focused in on it.
It stared back at me, though it didn't move.
I watched it for a long time before it finally stirred
And crept up to a place that might improve

Its chance of disappearing, of hiding from my view,
A place with bark and lichen in a groove.

First I thought I'd lost it; this time I couldn't see
If it was climbing or just sitting still,
When suddenly it left the tree and flew down
  toward the ground
But landed in a tree up on a hill
That wasn't far from where I sat; in fact, I saw
  it creep
Exactly like its name conveys its will.

From bottom to the top of a dead snag it worked
  its way,
Its tail supporting it with every rise.
Between each creep, it flattened out against the
  snag's smooth face.
It spiraled 'round as it moved up clockwise.
I guess it must have figured I was far enough away
So not a threat to commandeer its prize.

AMERICAN WHITE PELICAN
(Pelecanus erythrorhynchos)

A bird associated
With southern climes, not cold,
Is odd when seen up north.
It doesn't fit the mold.
I know it seems unlikely
That this bird's in this book.
But it comes every summer,
I swear. Just take a look.

My family lives in Florida.
I must be the black sheep,
'Cause I like colder weather.
Hot weather, they can keep.
Of course, they see the pelican
Whenever they go out
To visit gulf or ocean.
Down there, these birds have clout.

The pelican *does* migrate;
Great Lakes are where they drop
To take some needed downtime.
We offer a "truck stop."
They never seem to linger;
A few days, and they're gone.
I have to give them credit.
"Big white birds, carry on!"

NORTHERN FLICKER
yellow-shafted
    (Colaptes auratus)

Buff below;
Black spots evident.
Brown-gray back with black barring.

Chest patch black.
Cheeks sporting moustache.
Crown gray with red nape crescent.

Flash of white,
Fast-disappearing,
Found on rump when departing.

Underwings
Unobtrusively
Umber-spotted and golden.

Widely found
Woodland dwellers but
Winter-over south and west.

COOPER'S HAWK
(Accipiter cooperii)
NORTHERN GOSHAWK
(A. gentilis)
SHARP-SHINNED HAWK
(A. striatus)
BROAD-WINGED HAWK
(Buteo platypterus)
RED-TAILED HAWK
(B. jamaicensis)
ROUGH-LEGGED HAWK
(B. lagopus)
NORTHERN HARRIER
(Circus cyaneus)

I watch you in the summer,
    But in winter you move on
To climes that suit you better.
    And I know that you'll be gone
Until our snowbanks start to disappear.

When winter is behind me,
    Then again I'll start to watch
For harriers, low-flying,
    With a flash of white rump swatch.
And I'll be thrilled to see them reappear.

I'll look around the woodlands,
    The accipiters to see:
The Cooper's and the sharpies
    And goshawks—I love all three.
Each pilots like a trained flight engineer.

I'll scan high skies for buteos,
    They're daytime birds of prey.
Rounded wings and banded tails;
    Red-tailed—one band of grey.
Eyesight that is unbelievably clear.

And I'll listen for their calls,
    Each distinctly theirs alone.
Rough-legged gives a screech or squeal,
    Broad-winged "peee-teee" monotone.
The red-tailed does long and down-sliding "keeeer."

Then when summer days are gone,
    Once again you'll take the sky
On another lengthy flight.
    That's the time I say "Goodbye,"
But I'll see you when you come back next year.

COMMON MERGANSER
(*Mergus merganser*)
HOODED MERGANSER
(*Lophodytes cucullatus*)

Ducks with teeth? How can it be?
    They really are *not* teeth, you see,
But beaks cut out with pinking shears
    To grasp their food like little spears.
Hooded. Common. Cousins, you.
    You look somewhat alike, you two.
Male hooded heads are black and white;
    Commons—iridescent night.
Females' heads of reddish brown
    Can move crest feathers up and down.
You only raise one brood per year.
    Take caution when a threat is near!
Your voice, a croak, both harsh and low,
    Though in the spring you whistle so
To make a simple, plaintive call
    Among the kids, mom, dad, and all.
You're not so graceful on the land,

But you take matters well in hand
In rapids or the deepest pool.
 To you I say, "Mergansers rule!"

GREAT GRAY OWL
(Strix nebulosa)

Great is the perfect word to describe your
large size.
Resonant *whoo*'s, deeply uttered, comprise
your call.
Except in the far north of your range, you
are uncommon.
At night is when you prefer to hunt,
Though you also hunt at dawn and dusk sometimes
when you are farther north.

Gray feathers contribute to your name.
Roosting in a tree, you may be mobbed by
smaller birds.
As with other owls, you regurgitate pellets of fur
and bone.
Your yellow eyes look small within your heavily
ringed facial discs.

One time I saw you atop a telephone pole;
What a sight you were to behold!
Lacking ear tufts and having facial discs, you remind me of a barred owl, also seen here where I live (but they're not as great).

YELLOW-BELLIED SAPSUCKER
(Sphyrapicus varius)

My cedar trees are getting old.
Through summer's heat and winter's cold
They've stood their ground, erect and bold.
> But now, it seems, they've reached their
> golden years.

How do I know? How can I tell?
By oval holes, all parallel,
Where sap drains from a little cell…
> The birds no longer linger, it appears.

Sphyrapicus, who ate the sap
That drains after a brief cold snap,
Has ceased to stop and pause to lap
> The sustenance a cedar volunteers.

And now my trees no longer weep;
What little sap they make, they keep

To help them rest, to help them sleep.
    It's obvious they've ended their careers.

GOLDEN-CROWNED KINGLET
(*Regulus satrapa*)

RUBY-CROWNED KINGLET
(*R. calendula*)

I had a great experience
Involving these two birds
At WPBO* one spring.
(It wasn't just for nerds.)

It's the place I love to be
When they do their bird banding.
The information which they glean
Is far beyond outstanding.
Their practice is encouraged
By ornithologists
Throughout the whole United States
And Canada, as lists

Are paramount to following
The lives of avifauna
Around these parts in this great land.
And helping them—I'm gonna.

So anyway, on that fine morn
I happened to be present
When they were banding kinglets.
It certainly was pleasant
To be among the chosen few
Who got to help restrain
Those tiny birds, four inches small,
Though they did not complain.
Instead they trusted those of us
Who held them tenderly
As they were banded. And then
We were told to set them free.
I opened up my hands and whispered,
"Fly, now, little creature."

I held two kinglets on that day.
'Twas THE BEST double feature
In which I've shared the company
Of people just like me:
The ones who understand and feel
The joy of setting free
A tiny emissary that may
Make a difference

In how our world will prosper
And prolong its existence.

*WPBO = Whitefish Point Bird Observatory

PILEATED WOODPECKER
(Dryocopus pileatus)

Black and white
Undulates.
Flash of red,
Top of head.

Sudden stop.
Clings to trunk.

Drums: hammer
Strikes dead wood,
Makes large holes.

Wuk! Wuk! Wuk!
Calls to mate.

Uses nest
Only once.

Insect pests,
Like beetles,

Are its meal.

I always
Take time to
Stop and watch.

GRAY JAY
(Perisoreus canadensis)

"Robber" best describes you,
Snatching food from my picnic table
When my back is turned.

"Opportunist" you are,
Taking so easily all that you can;
Thus, it's not well-earned.

"Overgrown chickadee"
Some call you. It's still the same verdict;
Won't be overturned.

You have other names, too:
Whiskey Jack, also Canada Jay—
Those are two I've learned

From others who have been
Unfortunate victims of your stealth.
Nothing is returned.

It's hard to be angry,
You being so cute and so friendly
But so unconcerned

About all my labors
When preparing both food and table.
And I feel so spurned.

AMERICAN GOLDFINCH
(Carduelis tristis)

I think perky is the word
That describes this little bird.

Male is yellow, female's brown,
And she lacks her mate's black crown.

She shows hints of yellow, though,
On her cheeks and neck below.

Both have wings of black with white.
Both have white rumps flashed in flight.

They love Niger thistle seed,
So I make sure that I feed

It to them in socks of mesh.
I know that they like it fresh.

Like a rollercoaster ride
Through the air (that's how they glide).

Hear them twitter? Hear them pip?
Hear them say, "Potato chip?"

Late in summer they find mates.
Just the female incubates

Four to six small eggs, pale blue.
When they hatch, male feeds them, too.

At around two weeks chicks fly,
And the parents say goodbye

To their only brood that year.
For a while, though, they stay near.

Yellow male soon fades to brown.
They may stay or may leave town.

Flocks are maybe twenty birds
(Not big groups, in other words).

Some will seek the warmth and go;
Some will stay here through the snow.

I'll be here to give them seed,
Helping them in time of need.

SONG SPARROW
(Melospiza melodia)

Early summer, sun's out, sky's azure.
    Alone, sporting jeans and flannel shirt,
    I doze in my zero-gravity chair
    On the deck of my house
    Overlooking the bay,
    As ore carriers silently slip by
    Through the shipping channel
    Just this side of Canada.

I become aware that
    My solace is intertwined with
    The familiar sound of song sparrows.
    They seem to be singing
    For the sheer delight of doing so.

With eyes still closed, I smile.

FLYCATCHERS
(Tyrant and Empidonax)

In the U.P., where I live,
I've birded for countless years.
But when it comes to flycatchers,
I have to agree with my peers
That these birds are the toughest
To correctly identify.
I'm lucky to name one out of ten.
Let me clarify:
They're called, "The bane of birders,"
And that's right, I'll attest.
It's hard to name them just by sight;
They're similarly dressed.
Drab colors are predominant.
Bristles on their faces.
They have flat bills and large heads,
At least in several cases.
Sometimes I recognize them
By the places they appear,
And sometimes I can name them
From the voices that I hear.

For instance, there's the phoebe,
With its harsh emphatic call,
And the Eastern wood peewee
Says "pee-a-wee." That's all.
There's *Tyrannus tyrannus*,
The easiest for me.
It's called the Eastern kingbird,
And it's one I often see
When in a woodland clearing
Around a water source.
(That's a place not hard to find
Here where I live, of course.)
Some others I encounter
That share my U.P. home:
Great Crested, Olive-sided,
Least, and Alder when I roam
Around the bogs and in the swamps
And in coniferous woods.
I wish I could I.D. them
In my local neighborhoods.

CEDAR WAXWING
(Bombycilla cedrorum)

You are, without a doubt, one of my very
   fav'rite birds.
Everything about you is divine:
Your coloring, your soft, sweet voice, your crest,
   your shape and size.
In my book, you are nothing short of fine.

I love it when I see you frequenting my
   mountain ash,
The service berry, and the cedar, too.
I hear your high-pitched whistle trill, but I must be
   close by
To pick it up, or I just might miss you.

Your name fits you because, you see, it
   perfectly describes
The waxy patches you have on your wings.
They're mostly white. But also there are some red
   waxy dots.
And I will now describe some other things:

Your mask may hide your eyes, but it accentuates
   your crest.
You have a dark patch underneath your chin.
Your belly shows a bit of yellow, though it's pale,
   not bright.
And on your chest and head is light buckskin.

You are the most gregarious when winter
   rolls around,
Though you prefer some twelve-month company.
If ever you decide to leave your friends and
   social life,
I'd be so thrilled to keep you here with me.

GREAT BLUE HERON
(Ardea herodias)

Great blue heron
Barks like a dog
When he's startled.

It's the same size
As sandhill cranes;
Lacks their red crown.

Hunts for fish in
Shallow water.
Also savors
Insects and frogs,
Which he stabs with
Long yellow bill.

Conspicuous
Long black eyebrows
Which turn into
Dark feather plumes
Off back of head.
Feather necklace.

When he's in flight,
Neck's in "S" shape,
Legs are stretched out
Straight behind him.

Nest: a platform
In the treetops
Near the water
In colonies
Up to about
One hundred pairs.
One brood per year.
Both incubate
Three to five eggs
Which will hatch in
Twenty-seven,
Twenty-eight days
(Give or take one).
Then they will fledge
In fifty-six
To sixty days.

Migration is
Either to the
Southern U.S.
Or to Central
America.

AMERICAN BITTERN
(Botaurus lentiginosus)

There's always one bird that evades us,
The one we just can't seem to spot.
This one was my bane;
I couldn't attain
A sighting to fill up the slot

That gaped open there on my life list;
It glared back at me, sad to say.
So I made a pact
To finally act:
"That blank's being filled in today!"

I packed up my lunch and my binocs,
My trusty bird I.D. book, too.
I set out at five;
I felt so alive.
I'd have that bird 'fore I was through!

The N.W.R.* out in Seney
Was first on my plan of attack.

If there was a place
To see, face-to-face,
A bittern, then I was on track.

I crept along for three whole hours,
Searching each dense marsh and pool.
Then low and behold,
My plan did unfold:
There stood my long-sought-after jewel.

Its striped neck was stretched to the heavens.
It blended right in with the reeds.
It swayed side to side,
And I nearly cried
As I watched it there in the weeds.

But now I have more birds to conquer.
Blank spaces are many, it's true.
Though bittern is done—
And it was such fun—
I can't wait to see something new.

*National Wildlife Refuge

RED-BREASTED NUTHATCH
(*Sitta canadensis*)

WHITE-BREASTED NUTHATCH
(*S. carolinensis*)

Silly nuthatches—they live upside down.
One has a white breast, the other, red-brown.
White has an obvious black nape and crown.
Red has an eye-stripe (though not like a clown).
White makes an "ank" call; red's like a tin horn.
Both live in conifers when they are born.
Both of them like to eat nuts and some seed
Along with some insects. It's neat how they feed:
They wedge a large food item into a groove,
Then hack away at it. It's hard to improve
On such a good method; it gave them their name.
Both look really diff'rent, yet both act the same.

COMMON LOON
(*Gavia immer*)

Black and white bird of the lakes,
Checkerboard back and striped collar,
Does the color of your eye make you see red?
Do you view the world through
  rose-colored glasses?

You are the symbol of our northern wilderness.
Your name is from the Swedish word *lom*,
Meaning "lame," for the way you walk.
But when in the water, your grace makes up for any
  awkwardness,
Carrying your babies on you back, diving
  for sustenance.

You utter a laugh that, to many, suggests dementia,
Responsible for the phrase "crazy as a loon."
You are not crazy.
When I am sitting on the edge of a secluded lake
And I hear your laugh…
I share your sanity.

BELTED KINGFISHER
(Ceryle alcyon)

I cannot seem to help but chuckle at your silhouette,
Stocky trunk and heavy bill; you're perched on post or limb.
I can't mistake your small, short legs and boldly belted chest.
Your head is large with shaggy crest; your "hair" could use a trim!

You plunge headfirst to catch your prey: a minnow or crustacean.
You broadcast your existence with a chatter, loud and clear.
It echoes 'cross the distance, "I've completed my migration."
It's obvious you've claimed the parcel that you're fishing near.

I love the way you build your nest deep in a riverbank—
A tunnel sloping uphill so your chicks will not

get wet.
If flooding happens, they'll survive, and they'll have you to thank
For staying dry above the rising waterline. They're set.

You jointly do construction, working sunup to sundown.
I offer up great admiration for you and your mate.
You only rest to catch a meal. You work instead of clown.
When eggs are laid, your work's not done; both of you incubate.

And still to come is feeding all the chicks (there's five to eight)
By fishing nonstop all day long; they never cease their wanting.
Again you both work diligently, both participate
To guarantee that they won't starve. You're quite adept at hunting.

At last the chicks will fledge and leave the dark, warm home they've known.
But you must remain vigilant and now become their teachers.
Your work continues as you guide them to live on their own,

And one fall day I'll find you're gone; I'll miss you, little creatures.

BALTIMORE (NORTHERN) ORIOLE
(Icterus galbula)

Your bright orange body, your black and
  white wings,
The prettiest song that any bird sings—
I always rejoice upon your arrival.
I know it's been hard; I applaud your survival.
Your name was derived from a lord's coat-of-arms,
Which, when I see you, reflects in your charms.
The state bird of Maryland, mascot for their team
In baseball (for some, the American dream).
Your genus name, *Icterus*, is Ancient Greek.
It means "yellow bird." Your mate is most chic
With her golden plumage (though it's a bit duller
Than yours of bright orange, a flashier color).
You've come a long way with one purpose in mind:
To mate and have offspring; it's all predesigned.
Your genes and your instincts already contain
The efforts some might see as legerdemain.
You do your best singing from high canopies.

Your mate builds a nest in deciduous trees—
A pouch, tightly woven, that hangs from a branch.
You both feed the hatchlings. *You're* given
   carte blanche
To nest and to nestlings. But if things go bad,
You can't start again. And that's kind of sad;
You only get one shot at raising a brood.
And if you should fail, does defeat change
   your mood?
Or do you just take the whole outcome in stride?
I'd be most unhappy if I were your bride.
You seem to be unmoved; you just carry on,
And one day in mid-to-late summer, you're gone.
My feeders go untouched (except by the ants).
I can't help but wonder if some circumstance
Might have been avoided or even prevented,
Though you and your mate don't appear
   discontented.
All of *my* thoughts and concerns about life
Seem of no importance to you and your "wife."
Perhaps there's a message in all that you do.
Perhaps I should learn a good lesson from you:
I cannot control all that happens to me.
Take life as it comes and be much more carefree.
You have no awareness of thoughts you've
   awakened,
Of notions and insights you've actually shaken
Within my persona, deep inside my mind.

You've taught me to leave nasty troubles behind.
And now, on your trip back to warm,
   southern places,
You'll travel and not be aware of the faces
Of people who see you but haven't a clue
That you helped ease my struggle and helped me
   get through
A tough time in life; a time when I heeded
The lesson a bird taught me when it was needed.

EVENING GROSBEAK
(Coccothraustes vespertinus)

PINE GROSBEAK
(Pinicola enucleator)

ROSE-BREASTED GROSBEAK
(Pheucticus ludovicianus)

I don't know exactly the ounces per inch,
But they're stocky and plump with the bill of
  a finch
That's thick, and it's strong to crack open seeds
And fill up their throat pouch for their
  fledglings' needs.
Their flight's like most finches: they, too, undulate.
It's nothing they choose to do; it's just innate.
The males have their own special wardrobe—
  that's clear—
And no one can say that it's drab or it's drear.
A rose-colored triangle on the male's chest
Assures that the rose-breasted grosbeak is dressed

To impress the ladies so he can create
Some offspring that look just like him (or his mate).
The pine grosbeak's color is also distinct,
Bright red-orange and gray with white wing
  bars: succinct.
The male evening grosbeak—black, yellow,
  and white—
Has bright yellow eyebrows. He's truly a sight!
His bill's either ivory or a light green.
He's one of the handsomest I've ever seen.
The habits of each are all mostly the same,
Though each of the birds has a different name.
They all pick up small grains of sand by the road.
They also take snow baths soon after it's snowed!
Some years there are many and some very few.
I like the years when there are plenty. Don't you?

WARBLERS in spring

Golden-Winged (Vermivora chrysoptera)
Tennessee (V. peregrina)
Orange-Crowned (V. celata)
Nashville (V. ruficapilla)
Northern Parula (Parula americana)
Black-and-White (Mniotilta varia)
Blackburnian (Dendroica fusca)
Black-Throated Blue (D. caerulescens)
Cerulean (D. cerulea)
Chestnut-Sided (D. pensylvanica)
Cape May (D. tigrina)
Magnolia (D. magnolia)
Yellow-Rumped (D. coronata)
Black-Throated Green (D. virens)
Kirtland's (D. kirtlandii)
Prairie (D. discolor)
Bay-Breasted (D. castanea)
Blackpoll (D. striata)

Pine (*D. pinus*)
Palm (*D. palmarum*)
Yellow (*D. petechia*)
Mourning (*Opornis philadelphia*)
Connecticut (*O. agilis*)
Canada (*Wilsonia canadensis*)
Wilson's (*W. pusilla*)
Ovenbird (*Seiurus aurocapillus*)
Northern Waterthrush (*S. noveboracensis*)
American Redstart (*Setophaga ruticilla*)

In mid-to-late May,
Trees halfway leafed out and still very pale,
To the warblers I say,
"Just stop for a bit. I'd like to look at you!"

(Never mind trying to snap a photo.)

"At least be so kind as to sing for me;
I'll know who you are by the sound you make."

Sometimes, however,
It seems so hopeless and so frustrating.

"I think you enjoy trying my patience!"

BLUE JAY
(Cyanocitta cristata)

Although it matters not to you
    Just how you got your name,
The Greco-Roman languages
    Contribute to your fame:
Cyano, which is Greek for "blue,"
    And *kitta*, "chattering,"
Cristata, Latin word for "tuft"
    Describing your head dressing.
Combined, they make you "noisy bird"
    And one that's blue, to boot,
Who has a crest for all to see—
    Some even call it "cute!"

You serve a lot of other birds
    With early-warning call;
You scream when predators approach,
    And that's a sign to all
To hide themselves, get out of sight,
    There's danger to their nest
And all their eggs or baby birds.

They know it's not a "test."
If owls roost on a nearby branch
　　Throughout the daylight hours,
You and your mob will run them off.
　　They won't invade *your* towers!

You feed on acorns, arthropods,
　　And soft fruits, seeds, and nuts.
You're smart enough to hide excess
In holes and cracks and ruts.

You build your nest up in a tree,
　　And there you lay your clutch;
It numbers two to seven eggs.
　　You needn't worry much
About providing for your chicks
　　If there is lots of food.
Ma Nature's looking out for you;
　　She'll regulate your brood.

When all the young have left the nest
　　You'll still remain together
Until the fall, when young depart,
　　No longer on a tether.

TURKEY VULTURE
(Cathartes aura)

Circle o'er a fresh roadkill,
Ride the thermals, float at will.
While the eagles eat their fill,
    Ravens sit on branches and complain.

Pick the bones when all is still,
Bits of flesh—now mostly nil.
No blood's left for you to spill.
    Fur and antlers are all that remain.

Move on, past another hill,
Through a valley, over rill.
Focus on a different thrill,
    One that other scavengers disdain.

You prepare to test your skill.
Watch your step! Beware of quill!
Though, for you, it fits the bill;
    Porcupine—common in your domain.

COMMON RAVEN
(Corvus corax)

Common raven, bird of legends, "Trickster god" the
   natives call you.
Norse god, Odin, uses you as eyes and ears (*Huginn*
   and *Muninn*)
Bringing him the news of Midgard. Viking banners
   bear your portraits.
You appear in myths and folklore of a plethora
   of cultures.
As a messenger, you fly between two worlds—both
   dead and living.
Haida think you are Creator, bringing sun and moon
   and water.

You are represented throughout all the world
   in literature;
Bible, Talmud, and *Qur'an* all cite your presence on
   their pages.
Shakespeare oft made reference to you within his
   plays and sonnets.

Poe and Dickens emphasized your ominous and
   dark "persona."
Captive ravens of the London Tower are both loved
   and feared by
Those who trust the legend that if these birds leave,
   then Britain crumbles.

All your history escapes me as I watch your
   brave abandon:
Soaring, flying upside down, and doing air-bound
   acrobatics.
You are playful, smart and clever; you can mimic
   other species,
Solving problems, caching food, and learning, not
   just using instinct.
Profiting from human spoils: our garbage,
   roadkill, irrigation.
Walk with confidence, or swagger—even lope—
   across a highway.

How I love to hear your raucous voice, to listen to
   your croaking,
Watch you preening your black plumage, see your
   bristled, thick, and strong bill.
I know you will not fly south when winter brings us
   freezing weather;
I will see you waiting alongside a
   frozen-solid carcass

Till an eagle or a mammal opens it for you to
  dine on,
You, the opportunist, knowing you will eat if you
  are patient.

MALLARD
(Anas platyrhynchos)

"It's only a mallard."
I hear that a lot.
And I've even said it myself.
It's not that they're plain—
They surely are not!
But they're common, like "right off the shelf."

I guess commonality
Isn't so bad if
We make ourselves understand
That being abundant
Can be therapeutic;
Recognition is comfort, firsthand.

AMERICAN KESTREL
(Falco sparverius)

MERLIN
(F. columbarius)

PEREGRINE FALCON
(F. peregrinus)

Falco sparverius, American kestrel,
Most common and smallest up here.
I find it quite comical watching your tail bob
When perched, though I know *you're* sincere
About finding insects, small reptiles, or mammals.
Through all kinds of weather, you're near.
You don't seem to mind all the rain and the snow.
You survive even where it's austere.

On the other hand, merlin, I have to admit,
You confuse me when I must compare
Both you and the kestrel, though you are
   much darker.
If I see your tail, I can swear

Identification is truly spot-on.
If you're flying, however, when there is a glare,
And I cannot make out one single factor,
I have to abandon my stare.

Then there's the peregrine, black crown and nape
With black edge that extends below eyes.
If I'm close enough, I can spot you right off.
Good keys are your speed and your size.
You chiefly hunt birds, and you like open country;
In cities, you may urbanize.
But once in a while, when I'm near open water,
I glimpse you when you pierce the skies.

COMMON YELLOWTHROAT
(Geothlypis trichas)

Wildly clipped and crisp, I can hear it
Where I'm standing beside the river.
"Witchity, witchity, wichity,"

I know it's a common yellowthroat:
I see a flash of bright yellow, and
I can just make out the black facemask

As it flits among the dense thickets.
And I remember hearing this bird
At a place close to my sister's home;

She lives in Central Florida, and
She also likes to do bird watching.
(Seems like that runs in the family.)

There is something that fascinates me,
Though, about any bird breed that is
Taken "out of context" and isn't

Singing where I'm used to hearing it:
Same as people, birds have dialects!
Sure as shootin', I could make out a

Prominent slowness in the way that
Particular yellowthroat sang his
Primary song: it wasn't clipped and

Crisp, like the sound I'm used to hearing;
Clearly, it was slow, with a true drawl,
Capturing my attention. My mouth

Dropped open with awareness that I
Didn't consider thinking about
"Down south" birds "singing" much more slowly,

Elongating the syllables of
Ev'ry "word" they speak. I realized,
Even the *birds* "talk" slower down here!

MOURNING DOVE
(Zenaida macroura)

You're named for your mournful call—
owoo-woo-woo-woo.
I love hearing it when all
    my windows are up.
You are neither large nor small;
    Just about one foot.

Bob your head as you walk by.
    Build a nest of sticks.
You are one of the most shy
    Birds who visits me.
I wish I could edify
    That I won't hurt you.

There's a definite pink tint
    On your underparts.
I approach you, and you sprint
    Before you take off,
And your tail shows a white glint
    As you soar in flight.

Your wings whistle as you rise;
    Easily, you're spooked.
At my feeders, seek a prize:
    Seeds or fruit or grain.
You are common and comprise
    Much of what I watch.

RUFFED GROUSE
(Bonasa umbellus)

In one respect, you're after my own heart,
As both of us appear to love the snow.
You, however, take some extra steps
When winter's wind and flakes begin to blow.
First, you grow new feathers on your feet—
Bristle feathers, I believe they're called—
To act as snowshoes when you strut your stuff.
But unlike yours, *my* feet are feather-bald.
You dive into the snowbanks for your roost.
I, on the other hand, prefer a bed.
Throughout the year you sport a crown-like tuft
Of feathers, which stay raised upon your head.
You have tail feathers you can proudly fan
To help entice the ladies in the spring.
You stand atop a log and beat cupped wings
To make a drumming sound (since you don't sing).
You have black "ruffs" on both sides of your neck,
Thus, giving you a reason for your name.
In our deep woods you eat the buds of trees,
Although I sometimes see you on the street

(There must be things that suit your appetite),
But I'm aware that's not your normal beat.
When danger is a threat unto your young,
Your brave, smart female hustles to distract
The predator who might well steal your chicks.
She truly keeps your family intact.
I guess there isn't so much that we share,
You and me; that is, except for snow.
And when the winter temperature drops,
I think of you and what you undergo.

BLACK-CAPPED CHICKADEE
(Parus atricapillus)

Friendly and trusting,
You sit on my hand
When I offer up a treat.

Winter seems no threat.
Summer is your friend.
Always finding lots to eat.

Who could not find you
Cute and appealing,
Your appearance and your tweet?

Black, white, gray, and buff,
Your feathers are plain,
But their pattern is so neat.

"Chick-a-dee-dee-dee"
Or your "fee-bee" call,

Your communique, so sweet.

Love your company…
Having you around
Making ev'ry day complete.

PIED-BILLED GREBE
(Podilymbus podiceps)

Bill...
>Thick (like a chicken's).
>Black ring around it.

Eye...
>Very very dark,
>Each with a white ring.

Tail...
>White and feathery.
>Puffy underneath.

Size...
>Thirteen inches.
>Smallest that dives.

Feeding...
>Dives for crawdads, fish, and aquatic insects.
>Stays submerged for extremely long periods.

Safety…
> Sinks like a submarine.
> Surfaces far away.

Specialties…
> Short wings.
> Lobed toes.

Nest…
> Platform on floating mat,
> Built by male and female.

Brood…
> Five to seven.
> Just one per year.

Migrate…
> Mexico, Southern states, Central America;
> Complete migration with no overwintering.

Delightful.

WHITE-THROATED SPARROW
(Zonotrichia albicollis)

Folks should pay attention to these small and
   common guys.
(That's strictly my opinion but important in
   my eyes.)
They're widespread birds that are at home in
   any habitat.
Some remain well hidden, but there's one aristocrat.
As Edith Sitwell* put it, (I'll repeat it here, aloud)
They're "entirely uninfluenced by opinions…of
   the crowd."
They go about their business free of worry
   or concern.
They aren't aware they give me so much pleasure
   in return.
Just as their name suggests, they have an obvious
   white throat.
Their upper parts are rusty brown; that's nothing
   much of note.

But that is where the mundane stops and all the
  assets start,
The thing that makes this small bird shine—that sets
  this bird apart
From other little sparrows is: they wear a
  special hat—
A black and white-striped bonnet. But they don't
  stop at that.
There's also a dark eyeline, but that's not their
  best trait.
The thing that sets this bird apart (I won't
  exaggerate)
Is one bright yellow spot that sits between its eye
  and bill
On each side of its face. And I can never get my fill
Of watching it, hunched over, scratching for a seed
  or two
That's fallen from my feeder. That's just how they
  make do.
But most of all, I love the song it sings throughout
  the day:
"Oh, sweet Canada, Canada, Canada" sounds like
  what they say,
Though I'm aware that they do not speak any
  human language.
They just engage in "sparrow speak," and it's to
  their advantage,
As all the other sparrows of their kind know just

what's said.
That way they all communicate and will not
  be misled.

*British poet and critic

HORNED LARK
(Eremophila alpestris)

Tsee-ee, tsee-titi—
Your call may not be strong,
But your constitution is.

Horns made of feathers—
They may not win a fight,
But they win you a mate,

Black chest and stripe under eye—
You may not win a beauty prize,
But they make you distinct.

Dirt fields, gravel ridges, shores—
You may not have a glorious home,
But you thrive in all places.

You live in my U.P. year-round.

WILD TURKEY
(Meleagris gallopavo)

The male turkey,
Most impressive:
Iridescent,
Head is bare-skinned
(It's blue and pink),
Blackish breast tuft,
Bright red wattles,
Tail that fans out,
Legs with sharp spurs,
Low rump feathers
Tipped with chestnut.
Gobbling call is
Heard for one mile.
Female's duller,
May lack breast tuft.

They live in the
Open forest.
Forage mostly
On the bare ground—

Seeds, nuts, acorns
And some insects.
At night they will
Roost in thick trees.

They are seen more
Commonly now
Than in the past—
Often find them
Crossing roads and
Scratching along
Roadside ditches.
They will even
Come to feeders,
Though they tend to
Make great messes,
(And they leave a
Bad aroma).

KILLDEER
(Charadrius vociferus)

I walked along a shore on the edge of
 Lake Superior,
My eyes down, concentrating on the stones at
 my feet,
When I became aware of movement ahead of me.
I looked up to see a solitary killdeer
Which had evidently been traveling at the same
 speed as I
But had turned to face me simultaneously as I had
 looked up.
Its two black necklaces against its white chest
 glistened in the sunshine.

I looked just past it and could barely make out two
 yellow "cotton balls,"
Both frozen in time.
Since we were heading into late summer,
I figured they were what was left of the second
 brood of the year.

I frowned, thinking, *Though killdeer are classified as shore birds,*
They're more often found in a vacant field or along a railroad track,
Neither of which was close by.

The adult bird sized me up.
Then it turned,
Gave one *kill-deer* call,
And proceeded to scurry away from me, over the stones,
Chicks in tow.
It eventually turned toward a small dune covered in dune grass,
And it and the chicks disappeared.

I walked another fifty feet or so,
Then I turned to retrace my steps.
But again, I noticed movement ahead of me,
And again, I stopped and looked up.
Standing there, facing me, was a killdeer;
I could not tell if it was the same one or another.
At first, it feigned a broken wing
And appeared to be leading me toward the water;
When I didn't follow, it took flight,
Presenting its bright red-orange rump
And calling out its *kill-deer* cry over and over as it flew

Until it vanished at the point where sky and beach met,
Though its voice lingered a bit longer.